The Trinity

The Father, the Son, the Holy Spirit

By

Nigel Edward Shindler

CreateSpace

4900 LaCross Road

North Charleston SC 29406

Copyright 2000-2013 Create Space, a DBA of On-Demand Publishing, LLC

www.createspace.com

The Trinity, The Father, the Son, the Holy Spirit

by Nigel Edward Shindler

Published by CreateSpace 17/01/2013

ISBN -13: 978-1482012675

CONTENTS

The Trinity

Part 1

The Stories That Created My life

Heavens' Enchanted Journey

Part 2

Memories Of All Creation

And God said; " Let there be Light."

Part 3

God is Good

Man Be Kind

Part 1

The

Stories That Created

My Life

Heavens' Enchanted Journey

CONTENTS

Heaven's Enchanted Journey

FORWARD   PAGE 6

Poems

FORWARD

The

Miraculous Dream of

Life

The stories that fill our lives, encompass our days,

Create meaning, a sense of order, and dreams we hope one day to

Achieve.

When all is said and done,

Nothing can be added or taken away,

As if time had decided how long we stay.

In the beginning, or so we have been told, The Lord created the heavens and the earth. The earth didn't have a shape, or contain anything of any kind, and was devoid of light. The Lord surveyed all He is, and decided to use Himself in the best way He knew how; He created a story, and brought it to life. From the heavens He could view it, and also be a part of it.

Why would He do such a thing? Why do any of us do anything in this world?

When we are close to finishing a day, and someone asks us how our day went, what we did; how do we tend to respond? We simply make up a story that the listener can understand and appreciate, involving people, situations, and events, which at the same time reveals who we are, what we are about, how we perceive things, and on and on; why should the story of creation be any different?

Using the tools available to Him, the ethereal waters that suspend objects in the space created by light, He created a Show that serves the same purpose. The greatest quest any of us can engage in is the search to better understand ourselves, similarly, God decided to do much the same, using Man as His primary instrument.

Contrasts expose differences, they highlight strengths and weaknesses, separate the good from the bad, light from the dark, the greater the contrast, clearer are the features contained within each; by the end of the Tale it should be more apparent that at any time before what The Lord is, what He does, and what He is about.

The story is transmitted on streams of light, that later coalesce to form the Stage on which a Play takes place. Each man has a role to play so the production can come to a successful conclusion.

Man searches the heavens containing a multitude of stars, and wonders how on earth he came about, what purpose does he serve, where did he come from, why was he placed here? The answer to all these questions is revealed just as the Play is about to conclude – the grand finale, so to speak.

In the same way that light gathers to form something solid, man binds to his fellow man in order to create works that are grander in scope, greater in value, more enduring in quality; the connecting, joining, fusing, of ideas, helps to manifest the most wondrous, miraculous, events that constitute our lives.

Beauty lies in the eye of the beholder. That which is within you, is reflected upon the world that surrounds you, that which you see, and also take care of; is it then so strange that many of the most remarkable, lasting, enduring, beautiful, pieces of art, have stemmed from those who have faced enormous struggles, endured tremendous suffering, and overcome monumental hurdles, and during times that seem the hardest they could ever be?

There is a better place for us all! Ask practically any novelist what he considers his finest book, he will likely tell you it's his next.

Birds of a Feather
Flock Together

0. Birds scatter within the wind.

The leaves carry the tunes which they sing.

I have never before seen a sight

So full of heavens' delight!

Rays carrying beams of light

Spot the precious need we all have to diminish fright.

On the ground a hill arises, around which a stream flows,

All the while

A mild, mellow, wind blows.

Off in the distance

Mountain ranges bear tips covered with snow;

How could the good Lord do so much?

I can scarcely even begin to know!

Forever more, I will carefully watch the seeds I sow,

So that one day

I will join with those who appreciate most

This world that should be pictured embraced in a silk bow.

Clouds drift, a rainbow shifts,

Rain pours,

At the same time my spirit soars.

Beauty rests within the dear rich glades of the distant moors.

The eye opens, reaches, then clutches,

Bringing forth the pageantry to its own,

Making it seem like the caressing touch of my home.

All lies in the sight of the beholder,

As I get older may I become bolder and wiser,

Becoming like the prophets that existed long ago

It is truth alone I wish to know,

To tempestuous temptations,

I must say no.

I will never stop, rest, cease,

Till the answer to my quest arrives,

Forever grasping truths

As I head toward the place I wish to go.

River of My Heart

Snow melts,

And starts its passage to the sea.

Here, the incline is steep,

Rain, snow, and hail, emerges,

From those above who has cause to weep.

The water tumbles

As it hurdles over rocky terrain,

The creatures that reside in the sky

Notice the speed is great,

Here they watch, to drink they refrain.

Bubbling, chuckles of laughter,

This is the sound brought forth.

That which nourishes offers rivers of joy.

Collected on a plateau is a pool you have grown to know.

From branches above, into the water you dove,

Others watched, enjoying the show.

Children love to play.

The gift is to take in their pleasure.

We all are really still small,

Acting as if we have some sort of knowledge,

Nothing could be further from the truth.

Mysteries emanate when you arrive within a forest.

If only more had cause to speculate,

The world today would be in a much better state.

A line is crossed, water seeps down.

Easy! There's no need to frown,

A greater, longer, more challenging journey awaits.

Why not celebrate by wearing a special, heavenly, gown.

Meandering over rocks, and through fields,

Lush greenery appears on each side, and also far beyond.

From these patterns sights emerge

That many have grown fond.

Sheep, goats, and horse, graze,

Seemingly ceaselessly,

But still content with the manner of each day.

A man can build a boat, make sail, and then charter his own course,

Surveying the skies above,

While enjoying all the things there are to love.

A dream can be constructed many ways;

The key is having a destination,

And listening to the words a wise man says.

Aphorisms are kernels of truth.

A small number of words can contain wisdom,

But still escape those who choose to live their live trapped in a booth.

Knowledge that is practical and useful

Can be acquired by travelling to many lands,

Listening to all sorts of bands,

And also shaking with humility plenty of hands.

Words can be multifaceted, multilingual,

A meaning is harbored in each,

And viewed as a miracle that appears on a white, sandy, beach.

This is where water will eventually arrive.

An ocean is where fish and mammals thrive.

Beneath images can be murky,

But we can now imagine a lot.

Amazing is the sense of realization you've now got!

Each soul can dip into the water that stretches forth.

A tide is created by the moon, as well as a star to the north.

Look up! What do you see?

Golden lights have a tendency to create wondrous fascination.

One day all this will be revealed in a chosen nation.

Shalom, means peace, hello, and goodbye.

What really is the difference?

Hi, How, Sow;

What does it all mean?

Pray to the lord,

And offer a courteous bow.

Without His love you would not be here now.

High, Low, Giver, Taker

I have sung and danced on fluttering wings

My imagination has no limits that I can foresee

I am continually amazed by the wonders that each day brings.

By following the example of Christ,

A day will arrive when I will become all that a man can possibly be.

The place may not be predetermined,.

It could be a forest,

Or maybe the foot of a peer that stretches out over a sea.

When this time arrives life forever more will have no bounds,.

Like Ramakrishna,, I will become a Seer.

This is the height man was made to reach,

But, still today so many live merely to survive.

They have no knowledge of the pearl of wisdom

On which they can thrive.

When, with no fear in their hearts,, they dive into the ocean of life,

Shaped like a pyramid, some will rise and arrive at the top

While others will have to walk in the dirt and sand,

Continually treading on the treacherous terrain of the land.

My sole remaining wish is that more, eventually all,

Will choose to live as they should.

Heaven is where they are meant to be,

Not among the shrimp and crab that lie

At the bottom of the deep, blue, sea.

Ode To Joy

I have seen wonders,

I have seen pain,

I have seen light on the earth once again.

Oh, how enlightening is the wisdom you seek.

No cares, worries, bothers;

The kind words the birds chirp, is what you greet.

A rabbit shoots from a ravine,

A predator follows the sight in his eyes;

What knowledge from this can we glean?

The only thing chasing you

Are the desires you create in your mind.

All around you is so beautiful, rich, and dear,

Why not just sit back, relax, and enjoy a thick, dark, tasty, beer.

How many things do you really require?

The more you have, greater is the burden you shoulder.

The smaller your weight, swifter is your gait.

Mind, body, heart, and soul, become ever bolder.

The simplicity is stated in the fact,

You are becoming more That.

Mysteries enfold, entwine,

Creating illusions fuelled by the scenes that lie between.

You are complete.

That you have surely been.

Is it your shadow, or the future, the guides the present unseen?

Be here, be now, and believe in God.

The Kingdom is there within you,

Merely watch carefully the things you do.

You are man, not beast,

And you don't make the sound, Moooo!

Therefore, be as you are, the greatest thing by far.

Hallelujah! Now you can now celebrate,

Have a drink with your mates in a Bar.

Life isn't serious, but quite hilarious.

Enjoy your problems, as well as your troubles,

They can only serve to enrich.

Stay clear of those that are nefarious,

Then grand, fantastic, will be your time on earth.

Your days will be spectacular and marvelous,

Your heart will be overwhelmed with delight,

And nourished by a heavenly mirth.

Kindred Matters
To Many

Walking down a country road,

A sparrow fell upon his toes.

He could not hear laughter,

Because it was night,

But he felt angel near,

So he knew, there was no need to feel fear.

Or even the slightest bit of fright.

The path is long;

Should he sing a delightful song?

Shadows were there,

Though, life, all considered, is fair.

We have rich pleasures to seek.

Our greatest quest is to reach our own peek!

Glance around,

Flying doves fill the air,

Swirling, twirling, to heights unseen.

Doesn't that really say it all;!

Innocent, free, liberated,

You should always be.

If you like,

You could discover al this while walking beside the nearest sea.

Yoga And Me

Yoga, of the truest kind,

Is something that creates a splendid mind.

Thoughts rush, hush, suspend, and blend;

All serve to build a society that can be enormously kind.

Some now realize kundalini is the key.

Many enjoy its presence, including me.

The heavens are opened; all is seen as brilliant, bright, light.

Only the privileged have this gift,

To everything it offers a lift.

Others won't understand,

They'll treat it as a miff!

But, without a shadow of a doubt,

Those that experience, know what it is about.

Without exception one grows stronger with each passing day.

The mind, and so much more, function better,

Further in tune with the nature of things.

Mysteries open, including the nature of the moon,

And the cry that emanates from a loon.

God's grace is the hand that guides the plan.

He is not of use to us, we are at His disposal.

The reason this should be so, we really will never know.

All that is clear, is that things pass,

Because He is the one that holds us so very dear.

Whether it is day or night,

It is important to strive to always do

What you know is right!

Mistakes will be forgiven.

Charitable acts serve as compensation.

This one accepts always, without the slightest hesitation.

This is the underlying factor;

Simply do what is good,

Nothing else will ever matter.

God is gracious and kind;

Strive to be of like mind,

Then the world will be your oyster.

Gifts will flow like showers from the heavens above

Drenching you with delight.

Each and every soul,

Is a bright, fabulous, glorious, light!!

Life,

With All Its Faults

Is Pure And Perfect

The meaning of it all

Is contained within the essence of the mind.

Mysteries fulfill our desire to mystify.

Displayed are the wonders that drain the ability to terrify.

Forces of light and dark are present in time.

As seconds pass by,

Each perceives its passage in a different way.

For some it possesses no reality at all.

All we have, or will ever obtain,

Is here, now, in the moment.

The consequence is a heightening of awareness.

Channeled is the evacuation of the desire

To recognize an appraisal of fairness.

An individual never has anything more or less than he deserves or needs.

Happiness, contentment, if not bliss, is continually within your grasp.

Hold out your hand!

Let it rest to comfort on the shoulder of another.

Whether one can conceptualize, visualize,

All as one together we will eventually be.

Let no man fear that this will not occur.

The mystics, prophets, angels, seers,

Were provided to serve as God's lure.

Follow, listen, observe, pray,

Think how marvelous is this day!!

Hiding In Shadows

A planet is suspended in space.

From a star light is projected.

A curve is the shortest path to proceed,

To reach the destination it wishes to feed.

All may appear to happen by chance, but mystics, and more,

Through gleeful dance,

Have revealed that the entirety is planned.

Read well the revolutionary philosophers of France,

Their words reveal wisdom is mightier than the sword.

Fight for what if right!

Then will open whole new dimensions.

Invention requires much perspiration,

Inspiration, motivation, innovation, intuition;

They will eventually bring the lies of a Ford.

It was the first to move on four wheels.

Gas propelled its motion.

Similar to light,

Its nature seems due to a strange potion.

A day will arrive when all will know the supreme Truth.

Nothing at all is due to a throw of dice.

The Master, Universal Force, monitors, maintains an Orchestrated plan.

The beginning lies in treating others with a fair degree of couth!!

God's Helping Hand

Encircling, entwining, combining, gathering within,
These are the forces that take us
To where we begin.

The soul is our heart, where our essence resides,
From it we should seek to never hide.

We have all we require to be happy and content.
Wants and needs are often misspent.
Don't allow the world, which is quite bent,
To drive you to places from which you were not sent.

Focus, allows love to grow wide,
Humanity has a presence which is immense.
Many still will not lie by your side.

Goodness, grace, kindness, is what we are about.
Around is grime and dirt, despicable things that cause disease,
And make you holler, scream, and shout.

Remain as you are, a sentinel being, that others perceive as distant and far.

Suffering is due to ignorance.

The wisest know this, and thus, perceive things clearly.

Each one deserves to be loved dearly.

Practice!

Create the knowledge that builds faith.

Heaven sent, are beings that help build mighty temples.

A prayer, a song, displays a sense of a need to long.

Individuals we all are.

Special, supreme.

Riches to each one the Lord brings.

Hymns, and so much more, bring a sense of meaning.

All as one let us rejoice,

Hallelujah! Christ is on Earth.

Each day,

Create within yourself a new birth.

Time is ever present.

Nothing else really exists.

Allow these thoughts and feelings to persist,

Then when the moment is right you will realize

You have won the great fight!!

By The Sea Side

Sitting by the sea shore

I realize life should never be a bore.

Waves proceed, and then recede.

Sea shells lie in the sand.

Never before has my life appeared so grand!

Beauty is contained in one's sight.

An eye lash can produce a tear.

The slightest thing can produce our end.

Why, for heaven's sake, should we live in fear?

A garden is hidden in a forest.

Roses, dandelions, daisies,

Many occasionally have fairies, elves, wondrous things;

They are part shown, sometimes revealed, and also lie on leaves.

Dazzling sights greet the soul.

Stare in the water, the rituals include a bowl.

A wand is waved, signs are given, and energy springs forth,

In every direction, including the north.

A
Paradox

I'm tumbling as I stumble,

As I search for things to hold dear,

Mumbling words not even clear to me;

What cost am I paying for these things?

What could it possibly be?

In truth, there is no price, no fee.

Head over heels I fall.

Never ending troubles fill my day.

What more could I possibly say?

But still wonderments greet me as I make my way through this haze.

How pointless, meaningless, it seems.

Where could I possibly be headed?

Events seen and now passed, I have met,

They have offered apparently a glean, an insight, into things unseen.

Tired, weary, I end each day, having accomplished what?

I cannot say.

Here I am in the present, that really has always been enough,

Really, I am quite tough.

Lonely, alienated, apart,

So few seem to have a heart.

Why do they make themselves so dark?

My friends, and neighbours, see things in a similar light;

I rely on them.

Here, there are connections that mean a great deal to me.

How could I have thought otherwise?

My, I do at times have some gall!!

Through life, I think of the end.

Hideous, grotesque, this quest, but still, I continue on,

Long after I have gone,

Memories of loved ones tell me this is true;

But, they rest now with me.

I will, no doubt, do the same,

Whether I'm liked, hated, or blamed, for things done or not,

The point is, this is all I've got!

Ups, downs, things scarce and bare;

Great are those things that mean so much.

Sometimes I blush, thinking of things offered.

Again and again, they flow and gush.

Life is all too much,

I'll take it all the same!!!

DISCOVERY

Discover your being, find your heart,

Know your fellow man,

Though, he seems lost in the dark.

A passage is hidden; where does it lead?

It doesn't matter!

Wait till the light passes through,

Then you'll find it all leads to you.

Extrapolate, manipulate, cogitate, divide,

I'll find the solution when I have nothing left to hide.

Oh My!

A Butterfly

Between shadows

A light dances in the mist.

Birds twitter as they flirt with the wind.

Slithering snakes wave over mounds and hills.

Who over all this guides, wills?

Innumerable creatures inhabit the seas, lands, shores, and sky.

Collectively, they sing a tune heard by those in touch, a part, of it all.

On a knoll he searches what is above.

Clouds are within his grasp.

The power, the pleasure, is enough to make him laugh.

Is he different, separate, the same, or ethereal?

Reality is the treasure that holds the answer.

Gently gliding within a gust,

A butterfly battles to reach a hand,

Determined, it must be,

It has decided where it wants to land.

Down below,

A river meanders through pockets of fields and wooded patches.

The man awaits the freckled winged friend,

While he absorbs the graceful beauty contained below.

When, finally, a touch is made,

Another hand captures the insect.

Its life is slim, fast,

Before there is time to fade.

The magic is to savor within

The precious, luscious, truth while you can.

This alone, is what you require to be a man!

Holy Life

Jesus lived many years ago.

He offered hope, meaning, life;

But still to this day

So many say,

"So!"

Crowds gathered as

His words reached every heart.

Maybe not then, but later,

Each and every one hit its mark.

Many rushed to kiss His feet, and touch his flowing robe.

Stories are told about Him till this day

To peoples all around the globe.

Why should it be that one man could represent so much?

The wisdom offers hope that makes our troubles fade

To the mere whisper of a hush.

It is you!

That stretched to reach the answer.

Realizing our Spirit makes us whole;

Contained it can be in the water placed in a silver bowl.

Each and every one has a Holy Host,

Even the fields filled with hidden mice.

Remember the philosopher, Hegel;

He taught us the meaning of the word, Zeitgeist.

Author

What is an author?

Someone with something to say.

In words he tells a story worth being told.

Something entertaining, he strives to make it,

So that till the end, we remain in his fold.

Life, lived through characters, are there.

None reral, yet fascinating, so we stare.

Each has a role;

Like actors on a stage they play their part.

To do it well, they use their heart.

We find all in the end is made complete.

The pieces fit together so neat;

We cannot, thus, detect a single hole.

Past, present, future, may be used.

The best do so wisely.

Careful not to abuse, each one they nurse.

When they do not, to themselves they curse.

If all is done well,

They know their efforts might fill their purse;

But careful they must be not to strain too hard,

Or they will end up in a hearse.

Life spreads, expands, and takes us in every direction.

Some unseen, and hard to detect;

Of those some authors take notice.

Among the best today they might be.

Their quality we might label also being paranormal.

Call it the beyond

It may be up, down, here, there, everywhere, around,

But always it remains a great distance

From being understood.

Some are fearful of these realms.

What could be more obscene!

Don't fear the unknown;

Fear instead your inability to comprehend.

Uncover everything I say, till you reach the bone.

Along the way many skills will be honed,

Till one day the fruits of our labour will be obtained.

There are writers, who know some,

Maybe this is why they live in a large home!

Who Am I

Who am I?

I have many names.

They, by happenstance,

Reflect that I am a wave, a smile, then a courteous bye.

Greetings, longings, biddings, tidings;

All the while revealing, yet hiding.

Goodness, stoicism, impetuousness, commonality,

Patience, an essence of formality,

For many there might seem a duality.

Fascinating, illuminating, glamorous,

Many times a façade of utter banality.

Where beneath, behind, in front, is this figure within us all?

Many may wish to ask, but fearfully don't.

A gleaming eye, a playful smirk, dancing inside,

But, self-consciously, notes that he won't.

A sign within a hand indicates understanding,

During, while, all this takes place.

He, I am, is contained within.

Sensing a presence unknown that stretches forward.

The longest distance sets us back

To a time and place that all shall, will, did, begin;

Knowing what this entirely is the means to the key.

Many may not sense relevance, practicality,

However, somewhere is an essential faculty.

Isn't that wonderful!

A few have named Him such.

All of these kindred spirits really have much.

Meadow

The meadow of my heart,

Lies deep down in the valley,

From it, I would never dare part.

It holds many wonders you see.

Nature's glory is all around,

You can compare it to the flight of a bee.

                                    If you need help in understanding

                               These sumptuous sights that are all around,

                                        Just take my hand!

                    I'll guide you to a place

          Where freedom can be expressed to its' fullest grandeur.

                    Every thought, act, emotion,

                        Each potion poured,

                    Will be experienced as grander.

You will believe it is due to Grace.

Not yours, but the Other

That hides in His own far off place.

                    Appreciate all you are;

                    Far more human be far!

Now, let your vision rest on this heavenly acreage.

Your eyes may shift, stare, and glaze over with tears of joy.

Rejoice in your happiness,

You've discovered who you are.

Ay once mature, one with nature.

All of you.

Whole.

COMPLETE!

You Art Thou

A plain of waters contains depths below.

Waters were parted, sayeth the lord;

He alone knoweth the places they would go.

It represents, symbolizes, so many things.

Eventually these insights will be provided to all.

One day each man, woman, child, will be that tall.

The heavens, earth, sky, are god's creation.

He is all that is, was, will be, forever more.

Placed are delicacies that fulfill sight on every shore.

A sea shell, a tortoise, tad pole, jelly fish,

Mixed together provide an, oh so delightful dish.

That which we ingest in any way can cause no harm;

However, be cautious of what is said, spoken, done.

Question the dimensions they stem from.

Are they good?

If not, are they set to cause harm?

Fields, streams, wooden houses called country barns,

Are examples of how matters should truly be.

The heavens have allowed that one day this will

Surely be!

Blessed art thou oh Lord.

Never shall any of us go without.

Decide that this is something of which

You shall not doubt.

Knowledge is Truth;

Irrefutable, indefatigable, unconquerable, permanent,

He!!

I am becoming that, I am becoming.

He claimed this of Himself.

If you are to be so,

Place carefully a cross on a familiar, dainty, shelf.

The End

Is

A New Beginning

Part 2

Memories of all Creation

"And God said;

Let There Be Light"

CONTENTS

Memories of all Creation

Forward   page 50

Afterword

The Seven Deadly Sins   page 37

FORWARD

Memories of all

Creation

A man views himself in a mirror;

He can see himself as he is, as he was, and also as he would

Like himself to be.

Creation is there so The Lord can reflect on His nature.

It is a reminder to Himself

Of what He is.

Man was created in His image for this reason.

In the beginning God created the heavens and the earth, and the earth was without form, and void, and darkness was upon the face of the deep, and the spirit of God moved upon the face of the waters, and God said; "Let there be light."

There, in totality, is how creation came about, and why we are here on this earth.

The nature of God is exposed in the phrase above, which is from The Bible. He is much as the Hindu Holy Scriptures detail, separate from all creation, but has three entities, parts, of Himself, that he can use to His liking; Christians call this the Holy Trinity, The father, the Son, the Holy Spirit, or Holy Ghost.

The Father, The Creator, resides in the heavens, the Son, The Preserver, resides on earth, the Holy Spirit is the force that animates everything, and is, thus, The Destroyer, also called the Holy Ghost.

The light that tells the story of life, we are told, is good, meaning it creates, but all stories must come to an end, and this is why God reveals, exposes, to Himself all He is about at the time the story is brought to an end.

The breath of The Lord was placed within Adam, and at first he was much like The Lord Himself, created in His image, but was later given free choice and free will, which then distances Adam from his true nature; after all, The Lord is constant, everlasting, permanent, eternal, He does not change.

The task Adam must then undertake is to reach back to what he previously was, and was intended always to be. This task is complicated by the presence of Eve, who is manifested from a rib taken from the body of Adam.

Eve is drawn by sight and touch to the forbidden apple, and is unable to adequately question what she is told by the serpent, and succumbs to the overwhelming temptation to take a bite out of the apple; all though Eve has free choice and free will, her capacity to discern, discriminate, is lacking, and it could be speculated that this is due to a lack of use; much like everything else – practice makes perfect.

Adam makes the mistake of trusting Eve, not questioning where the apple has come from. Eve was created to be Adam's companion, to starve of a sense of loneliness, and thus brought about, or seemingly so, for his benefit, therefore, it is contrary to his understanding to speculate that she is one who can, or even has the capacity to, harm him.

The penalty they pay is being evicted from the so called, "Garden of Eden", and thrown, so to speak, into the wide world to fend for themselves. It is not long after in the book of Genesis that a murder takes place, the ultimate Sin.

Figuratively speaking, the more a man distances himself from the Garden of Eden, greater is the struggle he must face to return.

What is the greatest mistake, error in judgment, a person can make? It is to break any of the Ten Commandments, or to even question any one, this is, and should always be, considered something that can cause harm to oneself.

Today we live in a world where those who present themselves as the ones who can offer assistance, help, comfort, are actually most likely your greatest enemy, and a threat to your continuing existence; in another manner of speaking, everything is now the opposite of what it should be. There isn't any one of that has to look far to discover this truth for ourselves.

At the point in time all souls are lost, The Lord is once again alone, and the only thing left for Him to do is reflect on the story He created, and witness as a result the tremendous power he possesses, and the scope of how glorious He is. The story was created for His benefit alone.

Imagine what it is like for an artist who has worked long and hard on a painting, and when it's completed, has the opportunity to stand back and take it all in, so to speak. Everything that was included in the manufacture of the work of art, quite obviously, was previously within the artist. God was a master painter in the beginning; He just thought it would be nice to give Himself a reminder of that, and a pat on the back at the same time.

"Well done! You did a good job." He says to Himself. In turn he answers back, "I couldn't have said it better Myself!"

One World, One People

One Lifeline

The tide is coming,

Let it then take us away

To a place we can feel comfort and peace.

Heaven's there,

Maybe, it will intermittently appear?

An island can be a sanctuary, A continent, a place to explore.

The two may be parted by a sea,

But, in your heart,

If you choose, and try hard,

It can mean so much more.

Frequently

We stand, sit, and recline, among others,

But, our thoughts, memories, ideas, and reflections,

Remain solely within us.

This is the case

Whether one is in a classroom, library, studio, field,

Or as you transverse lands while planted on the seat of a bus.

The clasping of hands,

A hug, a caring glance, a playful wink,

And so many things more,

Can make us feel we are attached, connected, to those around,

But, really,

Reason tells us this logic is not sound.

We are forever, always, apart,

Even in the moments we are

Combined, intertwined, sharing juices as precious, and sweet as wine.

Seated at a table

People can be at each end, side, or borders,

As they dine;

Cups, plates, bowls, handed to the closest,

Or maybe the furthest from oneself:

Again, sadly,

Despite all this we are still solitary, completely alone.

To starve feeling of alienation,

Think things through;

The solution has no complexity to its configuration.

Say to each and everyone, "Good health!"

A smile will then gladly appear on each face,

Revealing joy, and great satisfaction.

The gifts you bought

Remain wrapped in a bag beside the front door.

Their value has been suddenly, but explicably, decreased.

Place them instead

In the cracked, weak, and hurting, hands, of the poor.

Everyone has, and knows,

They've been granted something grand.

Far greater than some material, substance, or object.

A crown is now imagined, fantasized, to rest on each hewad.

Everybody can now view himself

As a worthy royal subject

One would think the harmony that has arisen is familial.

There is a sense of contentment, and being happy.

A mother remembers to change a soiled nappy.

A father rises, leaves the table,

And calls from a desk phone his children.

Even this late in life they continue to mean so much,

Though, the skin is wrinkled,

And aches so easily settle into the bone.

Don't kid yourself!

The differences among us are merely an illusion.

We are all one family!!

1984

1984,

Was it a novel, a year?

Or a time to face the truth?

Man had chosen the wrong path,

The end had become near.

Ever since man has been corrupted.

This has been the case with each passing year.

The evident consequence is an increase in fear.

The horrors that fill the world are so many,

But few are mentioned.

Why bother when you have a comfortable pension?

Computers have brought this destruction.

Patience has been lost.

Man acts through illogical compulsion.

The focus to do "good" has been diminished.

My God! It's over.

We're all finished!

What more is there to say?

A dear price it is now we now pay.

Increasingly we lose the potential to feel gay.

So few are aware of how total is our nightmare.

Those that are number so few;

They are extremely rare.

I am an Island

Each man is an island.

We are born alone, live alone, and we die alone.

How do we bare the solitude of this condition?

Most often by having a relationship, or an affair;

Then we say to ourselves, things are now quite fair.

This you openly declare by saying,

I have a wife now, her name is Clair.

Really, though, all remains the same;

Clair is the one you've chosen to make living alone easier.

I'm not a "fibber",

Everyone knows this is true.

The reality of life

Is that much is about suffering.

We experience it, witness it,

But because so few make efforts to dismiss it,

Time stumbles as it tumbles around us,

Signaling that things are far from right;

In fact, that everything has gone terribly wrong.

The island is connected to a sea,

But neither is contained in the essence of the other.

The water surrounds the earth, but it is not part of it.

We need others to survive,

In a shell, however, we hide,

But we always know where we are;

Obviously, not in a bar.

Tell people the truth,

With a twinkle in your eye say,

"On a star"

A Good Man

A good man has suffered; what should his reward be?

Brilliant he once was, kind to all,

This was evident to see.

Day and night he read.

On words of knowledge he fed,

While others would dance and sing,

And speak frivolously about things that had once been.

I sat and examined it all,

Carefully using judgment to discern truth

From that which is false.

Ass days passed the heavens grew ever closer.

More and more,

He'd say,

Wisdom can never have too high a place.

A day's end he'd shut a book,

And through a window he'd look.

The stars would shine, the moon glow,

And think how remarkable it is

The extent to which he'd grown to know.

Long Lost Home

How many times did I stare out at the dark night sky, and wonder;

How on earth did it all happen?

Why?!

All hope would seem lost.

Carried away by a careless sea,

Crashing upon a distant shore,

Then finally ceasing to be.

The sound of a wolf crying in the wind,

Reflected the extent to which I'd lost the will to sing.

Encroaching from seemingly every side

Were the deeds of others I'd never before thought possible.

The only road left, I was sure, would lead me to a hospital.

Shattering upon a grimy floor,

We're the dreams, hopes, expectations,

I'd long ago believed would carry me to the heavens.

Projected along a path that would insure I would continually soar.

The strain was evident upon my weary, tired, face.

Their faces would stare with eyes containing pupils filled with mace.

Miles I walked, seeking an answer

To solve the riddle that would explain my pain;

Regardless of the amount of expended exertion,

There would be no detectable gain.

Somehow, throughout it all,

I stood tall, and managed to remain sane.

The excuses used by others to explain their lack of concern,

Always struck me as being quite lame;

Each was pitiful, and caused me to feel great disdain.

Their primary concern was merely themselves,

And acquiring a degree of fame.

All as one, the masses lost interest

In the one who yearned to provide others so much.

Any appeal for help, assistance,

Would be answered with fingers placed on lips.,

And the sound of the word, hush.

Had I ceased, and was made to die,

Others might carry my body to a grave

Upon which a tombstone might lie.

How many would attend this event to mark a life that has passed?

I wonder how many would remember the soul beneath?

Probably not many, maybe none.

The work, strain, effort, would leave no lasting impression

On the hearts and minds of those I'd struggled to care for.

No scriptures, books, letters, fables, would tell history;

He wasn't considered among the ancients of lore.

The night has now passed.

The days have grown old.

Rags are all that are left to cover, protect, each bone.

Across my back, and upon my shoulder,

Lays the hand of the Lord.

I have arrived.

Peace is here.

I am home.

DARK

WHAT IS THE DIFFERENCE BETWEEN GOOD AND BAD?

By far, it's not hard to say,

The difference is like night and day.

Light opposes dark.

Beware of the night, it is said,

Strange things hide, in shadows they bide.

Leave them be,

Then what you'll see will be as it has been,

Free!

Light energized.

Night dulls.

In this lull, the mind realizes null;

Thus, beware of the night, it's where there is no light.

It causes fright.

Leave it,

Or face a fight!

RIGHT AND WRONG

PEACE WILL COME

WHEN OUR WORK IS DONE.

COMPLETION IS WHAT MAKES US WHOLE.

DON'T CONSIDER IT WORK,

JUST MIGHTY FUN.

IF FROM DARKNESS WE ALWAYS FLEE,

THEN HEAVEN IS WHERE WE WILL BE.

I KNOW I'M RIGHT

WHEN I FOLLOW THE PATH OF GOOD.

MISUNDERSTOOD BY PRACTICALLY ALL,

I STILL MANAGE TO STAND TALL.

IGNORANCE IS BLISS

TO THOSE WHO DO WRONG.

WITH GRERAT HASTE

WE SHOULD AVOID SUCH HUMAN WASTE.

WE SHOULD PONER

WHY THERE ARE SO MANY THAT SQUANDER.

THEY SEEM TO HAVE ,

BUT WHAT THEY HAVE

WILL LAST JUST A LITTLE LONGER.

THE THINGS I LONG FOR

WILL REAMAIN FOREVR, PERMANENT.

A PLACE TO STAND FOR ALL ETERNITY.

IF THERE IS A TRRINITY,

I STILL CAN REACH INFINITY.

Evil By Chance Or Happenstance

What will evil do?

Whatever it likes, whenever it chooses,

That is its ultimate goal.

Its aim is that all deserving will one day live in a hole,

Where rats will bite till the bone it reaches.

No longer are these people living on glorious beaches.

By having so much,

They have insured that others have too little.

Things must always be to their liking.

They can be very fickle.

All this to them is a laugh,

As delightful as a playful tickle.

Caring for others is of no interest to them.

People die, children become orphans,

So that on their finger they may wear a colorful gem.

Wives must have dresses with an appealing, enticing, dangling, hem.

Who, in fact, has stolen more from the earth, women or men?

The end reveals the truth.

Life will return to how it once was.

The story of Genesis reveals the cause for our loss.

The serpent made a choice based on the knowledge it held.

Eve took a bite out of the apple that represents life,

And so began the beginning of our woes.

Why do the mystics give warnings?

Ramakrishna said, avoid women and gold.

They seem appealing, and serve so well.

But, ask yourself;

How many today have lost their soul?

It was just another product to be sold!

<u>MAN</u>

What makes a man?

Many things of a kind we do not understand.

Days pass, we are awake,

And when we close our eyes, we sleep.

Through it all we have a journey we must follow.

The answer to it all, we always hope, will happen in the morrow.

He search, we follow, until we can

Find shores of solid ground we can call our own.

Some call this our final resting place,

Others consider it home.

No matter how far, or how fast, we travel,

No time has passed.

We will always end up where we are now.

The beginning is the end.

The question many ask is; how?

Trying to understand all this

Can sometimes send us around the bend,

But still we pursue the quest to understand;

Though, from one day to the next,

We appear no closer to the truth,

To the Promised Land.

Could it be we need a helping hand?

The search without will make us weary.

Finally, we search within.

We develop a new theory.

The ultimate is obtainable.

True knowledge of everything that has been,

And everything we could possibly see.

The complexity of ourselves,

Is daunting to comprehend.

Layers, stages, rungs, stepping stones;

The essence of our being goes deep into our bones.

There, we can say,

Everything outside is a reflection of what we have within.

What is it we avoid most?

Is it fear?

It inhabits unknown territories guarded by a shadowy host.

We fear ourselves;

The truth that lies within.

We should ask ourselves;

Can we stand the truth, or will we fall, and never rise again?

Oh, the pain!

Existence is merely a game.

Let's end it,

And free ourselves from all suffering,

For all time.

The Buddha said it all.

Many Giants there have been,

But he stands as the most tall.

Consciousness

To be conscious, means, to be aware.

The degree to which others are,

I can hardly bare!

Given the chance,

The teachings of the masters will make life a dance.

Joyous will every moment be,

Because it is solely the moment you see.

Stage upon stage,

Are left for you to experience, discover, and explore;

But remain calm and secure while to new

Elevations you soar:

Because all of life is merely a show.

This, you should surely always know.

Actors we may be,

But higher and mightier is He.

To recognize how unique and special we are,

Reveals so clearly above all else we are by far.

Once faculties grow,

To new heights we eagerly go;

Leaving behind forever more any sense that life is a bore.

WOMAN

Where is it I wish to reside?

So I can recover from all the lies I've been told?

I want somewhere pleasant,

So I can consider it made of gold.

So often my skies have turned blue, then grey,

And finally to a state of black.

Stormy clouds gather then disperse.

So often these things have been due

To the creatures who prefer to carry a purse.

At times I like, in their face, spit a curse;

But, really, they are so spiteful and mean,

They'd probably strike aback, hard and fast.

Regardless of the pain I suffer,

They will consider fit so they deem.

Acts of kindness mask a horrendous darkness.

Fangs appear, stick into your side.

With a smile they apologize; all the while having a smirk inside.

These being were once thought of as caring, comforting,

Made of sugar, spice, and all things nice;

But, truth be told,

Your life they will surrender.

It's just a game with each throw of the dice.

If they tend to your hair,

Be careful and mindful to search after,

To your shock and surprise,

You may find embedded ugly, slimy lice.

They infect you with nonsensical sayings and words.

My God!

If they would just go away,

I'd be free to enjoy each day.

The
Will To Care

What does it mean to be thoughtful?

To use the mind, so you can help,

And thus be kind.

Still, there is confusion as to what this means;

Obviously, a great deal more than eating beans!

So much is required

To keep oneself healthy, stable, and strong.

Use to the fullest the resources you have,

Then, the better of we will all be.

Surely, this is something even a blind man can see!

How did it come about?

Destruction, despair, un-relinquished care,

Has made so much bare.

Does it cause a scare?

It should,

But so few care, or are even aware.

All of this is due really to having an inability to care.

Our days on earth are now numbered.

One day so many will die,

Then thunder will rule,

Lightning balls will shock them to life, And then finally with the devil himself they will have to deal.

So much could have been undone,

If only people had not loved fuel.

What does it mean?

Materials, convenience, a perceived appeal,

Something pleasant to touch and feel.

Sense rule,

While the heart struggles to beat.

So much is shown

By those who on a seat put their feet.

A lack of awareness, concern,

A willingness to please only the self,

While consequently others suffer as they burn.

Do they know of such a thing?

Atrocities occur

While in churches congregations choose hymns to sing.

A polar bear lies dead among a flock of geese.

Its' prey in not far away,

But, obviously, it has found its last day.

How sad this is to say.

CONTRAST

Where do I go?

That is what I wish to know.

The future is uncertain,

Everything is new.

What is it I should do?

Unanswered questions often

Leave me feeling blue.

Many ladders climbed, rivers seen, crossed and parted,

Suffering is plentiful, often I'm broken-hearted.

There lies the truth.

Good appears when contrasted with the opposite.

The world we need to explore.

Leave what remains in a booth.

Dark tides we face each day.

Once they recede a beach will appear.

Now can be seen the gifts left behind.

Pools contain creatures of a breed,

Birds circle above, searching for prey.

My God!

Here they are each day.

Is it knowledge, all these things?

The ebb and flow of life it will bring.

Enhance, embrace,

Ask for grace.

Let us all pray!

The Seven Deadly Sins

What does it mean when it is said that a sin is deadly?

What is it that dies? People commit sins while they are alive, therefore, it can be taken for granted that it isn't the life itself that ceases, but something else. What is that?

The answer is simple to provide – the soul; that is what actually makes us human. Without a soul, a person can function, but cannot live, which means being conscious, aware, and thoughtful, of that which surrounds you.

When the soul is weakened, a person's capacity to recognize the environment he is placed within is dimmed, and, therefore, he is less able to detect the manner in which his actions affect any milieu. The result that arises from this taking place too long is that the person may live in a pile of horse manure, but have no awareness of this, how the horse manure got there, and how it can have affect himself – health and well-being.

The world is hopelessly out of control, death, destruction, and despair, are quit literally everywhere, yet so many, remarkably, are unaware of how hideous our present condition is, and furthermore, how their actions have contributed to this overall state of affairs.

 How could such a thing be possible? One might ask; especially given the fact that history has shown that man has the capacity to do so much, persevere horrendous ordeals, and overcome tremendous hardships?

The key to formulating a clear understanding of these affairs is considering the meaning of the word, pride, and how such a thing can be a sin; it doesn't have any relation to the word, proud.

A person can be very pleased with his accomplishments, which can motivate him to continue doing more, but when pride becomes entangled in any work process, at some point someone will pay a price for the actions undertaken.

To give an example; two artists sit side by side at a table, each has a set of crayons and pencils, and both decide to create a piece of art. One of the artists develops a picture that attracts a lot of attention, and praise; far more that the artist seated beside him.

He learns to love the attention, and praise, he's given, and as a consequence, he decides one day to arrive earlier than his fellow artist at the table on which the two sets of art materials are placed in order to take a few crayons and pencils that are not his own. When the other artist arrives, he declares that he has no idea where the missing art materials are, and certainly had nothing to do with them being taken.

The "art thief" has decided to steal and be deceitful because he believes with better materials at his disposal he'll be able to produce finer pieces of art, leading to, he hopes, even more attention and praise than he received before.

He becomes so consumed with wanting attention, as well as praise, that he fails to notice the toll the other artist is paying. Later when he is reminded of this by something, he rationalizes, and justifies, his actions, by deciding in his own mind that he is more important than the other artist, and, thus, deserving of more; after all, the extra attention he's getting only proves this point - as far as he's concerned anyway.

It doesn't take long before the other artist cannot complete a piece of art because the one who has become famous has taken all the art materials for himself; as a consequence, art lovers are left with only a single artist they can rely on when they decide they want to view some form of art.

The artist that has become a celebrity believes he is actually great at what he does, without at the same time being able to acknowledge that he is no longer competing against another artist.

Viewing the evolution of events in the manner I have described, it is easy to conclude that being proud of work you've done, is fine, but it is imperative to recognize that every artist has an equal right to create, and the capacity to do so should not be hampered by being deprived of materials because someone believes himself more deserving or entitled.

The worst state of affairs imaginable is when a person loses his life because another has concluded he is deserving of the right to do so. It is most evident at that point that all seven of the deadly sins are deeply ingrained in the person's life, though, efforts to disguise such will also be present.

Try asking your neighbor, an acquaintance, a co-worker, how they acquired various things, accomplished certain tasks; you might get a sketchy, hazy, answer, that doesn't really make a lot of sense. When that happens take it for granted that not much sense is there, the soul is gone. All that is left is a person functioning in a manner that serves the purpose of fulfilling a role; absent is a human being who can evolve, and one day realize his Spirit, which is his true nature.

Heaven's Beginning In Time

Part 3

God Is Good

Man Be Kind

CONTENTS

Man Be Kind

Forward   page 91

Afterword   page 125

FORWARD

God is Good

The scope of our imagination can conceive of practically anything.

Strange it is then that we have destroyed so much of creation.

If we can put our hearts and minds together,

Maybe a new world can be brought out of nothing.

"God said; "Let there be light". He then claimed it was, good. In order to evaluate whether something is good or bad, it is important to determine what the purpose of the endeavor is; if it manages to accomplish the function it was made to serve, only then can it be concluded that it is good – of course, one is then required to determine what the word, good, means in the context I have described.

The word, "good", as I have used it above, represents creation, that which creates; as we can all appreciate, the better each of us understands how something operates, better able are we to use it proficiently; accepting what I have illustrated, creation is good because it enables The Creator to appreciate the nature of Himself.

How can anyone of us understand, appreciate, the nature of experiencing happiness without at some point experiencing sadness, disappointment, or dismay? In a similar manner, by viewing the nature of that which destroys, the nature of goodness, creation, can be more fully realized.

When the world is on the brink of complete annihilation, The Lord will then view the contrast between that and Himself, and have the opportunity to fully appreciate the nature of that which is good, which is of course, Himself.

A famous saying is, "Ignorance is mankind's greatest enemy"; and it is definitely accurate. If anyone ignores, doesn't pay attention, is not aware, of something that can harm them, it will eventually destroy the person that is not conscious of its existence.

The world presently is in a horrendous condition, and the primary agency responsible for this undertaking in the creature commonly referred to as being, human. He has incorporated within the earth's ecological system ingredients that are foreign, meaning destructive, and at a rate which enabled an accumulation to develop, which has resulted in a decay of the earth's ability to maintain itself in the fashion it is supposed to.

To make a comparison; a car may be constructed to run solely, or ideally, on diesel fuel. If another type of agency is used to operate the car other than what it is ideally suited for, its ability to operate efficiently will decrease, and increasingly more so over time.

How does a human being allow such a thing to happen to himself? Our nature, after all, instructs us to survive, and to struggle at all times to do so.

A human's greatest defense mechanism, that which protects his ability to sustain his existence, is the capacity to freely make choices, and a will that enables a person to act upon those choices. If either, or both, encounter a diminishment in strength, the inevitable result is a depletion in the person's ability to remain alive; similar to what has happened to the earth, when a foreign agency is incorporated within a human, resulting in damage to the faculties responsible for it to function well, it will correspondingly be weakened, and increasingly more so over time.

How does such a thing happen, and at such a rate that at some point a person may act in a manner entirely contrary to his nature? When such a state arises, quite obviously, the faculties of free choice and free will would be radically reduced in strength - for all intents and purposes, non-existent.

It is important to consider that normally when someone becomes cognizant of something that depletes hiss ability to function, he will take measures to insure it doesn't in the future infiltrate itself in his life; accepting the logic detailed, it can be determined that for a person to function in a manner entirely against his nature, the transformation would have to occur over a relatively short period of time.

The nature of most of the elements that make up a human being is that the more it is used the stronger it will become, therefore, one can conclude that anything that restrains an individual's capacity to freely choose, and take action upon choices made, is foreign to its nature. On a daily basis, this is exactly what people are doing to themselves; getting involved in activities, and ingesting elements, that weaken their most essential faculties.

Many foods that are commonly sold, and easily accessible, decrease the body's ability to function well. If a product claims to include an ingredient that acts in a certain manner, or serve a certain purpose, that it doesn't, for instance, something might be advertised as being healthy, beneficial to your health, but the truth is quite the opposite, the body will operate less efficiently as a result. There are not a lot of people today who are even close to their prime weight, being either too thin or too heavy. Most people, as well, are not as physically active as they should be resulting in muscular atrophy; many have resorted to using products that artificially induce muscle growth, which would be similarly destructive to a person.

The mind, much like the muscles in the body, needs to be kept active in order to stay strong. A person should always be aware of such a thing, and feel unrest, a sense of unease, when not being adequately stimulated, resulting in the person doing something to "occupy the mind"; therefore, it can be concluded, that if a person is deprived of the sense of such things, the person will pay a price, and the cost will increase over time.

Anything that dims awareness would, and should, be considered foreign to the person's nature. Activities such as smoking, excessive alcohol consumption, ingesting drugs, behavioral addictions, or simply being placed in an environment that has a hypnotic effect, such as sitting in front of a TV or computer screen for an excessive period of time, serve to weaken a person's will and powers of discernment.

A person might fall into a hypnotic state, and remain in such a condition for an extended period of time, but when such an episode concludes, a person should be aware of the detrimental impact it has had, and avoid doing such a thing in the future; if, however, the materials used to induce a hypnotic state are present without a person's awareness, the person is vulnerable to being harmed without knowing how or why it happened. Computer screens are practically everywhere today, and far more often than not, they are not necessary; they are in libraries, restaurants, bars, classrooms, university lecture halls, and on and on.

The very best way anyone of us can take care of ourselves, is to be moral, just, and honest, this is what keeps us aligned with the "order of the universe", and, thus, connected to the mightiest resource of all; by engaging in any activity that is immoral, unfair, or unjustified, we not only harm the person who is the target of these actions, but ourselves as well.

In much the same manner deceit is used to place a foreign, destructive, object in a person's life, we deceive ourselves by not acknowledged responsibility for what has happened, which is due to our not informing ourselves adequately prior to encountering the foreign object.

The end result, literally, is that a person can be entirely unaware of a massive loss of human life that is associated with his actions, thereby making the person, by definition, a mass murderer, and conceive himself at the same time as being moral, just, and righteous, all the while being unaware, as well, that he is the same as the monsters he watches practically every day on his TV and computer screen.

A PRESENCE BEYOND COMPARE

When in the heights of dismay,

The pleasure is to seek what we may.

Guarded thoughts hide dark feelings; surrounded we are by ugly buildings.

So many things can make us sink, stretched tight, almost to our brink.

Ups, downs, making plans to heal,

One day soon will be proper deal.

How can so much happen

That makes us question our continuing existence?

Is life fair? So many choose to be square!

Pegs to fill preconceived holes.

Choose a partner, so you become a pair.

Why?

Because it is told that this is what is supposed to happen.

It is the accepted pattern.

Break free! Be wise, learn for yourself;

The tools happen to be everywhere.\

Choose the time, provided is the space.

Life, the world, is an enchanting wonderment.

Fulfill all your dreams by cherishing the essence that is all of you.

Each one of us

Is superior by far from what we presently are.

Comfort, leisure, happiness, will fall into place,

When it is accepted that all will occur in its own proper pace.

You decide whether life is beautiful and kind;

There's no shortage of exquisite sights, things to savor.

Appreciate offerings, relish each and every flavor.

God has the answer to all our queries.

If you continually do what is right,

These will appear as an ongoing series.

LEARN ABOUT LIFE

Teach your children well, I say;

Then given to each will be a wonderful day.

Mistakes hurt, pain can make you strong.

Truths, knowledge, are the things for which you should long!

Learning can occur in so many ways.

Given time, less will be the haze.

Life, people, actions, become clearer;

Overall, life will become dearer.

So much is involved;

Parents, schools, pools, fields, and song.

Enjoy each;

They have something to offer of worth.

Therefore, rejoice each precious birth.

The son and daughter, have hearts that glow.

Minds create words, deeds to sow,

All going well,

That which you reap will be something deep;

And all one day will value and desire to keep!

ALL IS NOT LOST

What is bright, that we can see?

Stars shine, the moon glows,

Within is God, The Creator, this clearly shows.

Despite wars, fights started over nations, life continues.

Seeds will sprout,

Of this, never should you doubt.

We think things are so;

More should be offered.

This man states proudly, and this does he know.

Absolutes are absurd.

Truth lies buried in mists and streams.

Nature is our true call.

Leaves, branches, trees,

Plants, seeds, worms and deer.

Why wonder what life all means;

Live life to the fullest,

Shed every last drop of fear!

BATTLES ARE WON
AND LOST

A soldier holds a gun, a spear, afterwards, a beer,

He dampens what has happened with alcohol, and a long gulp;

Sometimes that is all that's left that can offer hope.

Blood, gore, seep from open wounds.

Cries of anguish, pain, and despair,

Fill the open, but stained, air.

Left behind are souls that are now lost.

Heavens appear, but choose not to grant release;

Instead appears the devilish beast!

Howls of laughter greet each death.

Chains will hold them in blistering flames of fire.

Their fate is now dire.

Never again will they perceive joy.

All hope will be lost, every sinner is included;

Jews, Hindus, every sort of Gow.

Now they will get what they are due.

No time is left to file and sue.

THE POWER TO
LEARN

A rush to judgment is what keeps learning at bay;

No matter the number of experiences, or times one can say.

Think, reflect, and discern truth from fiction,

Then gradually knowledge will occur.

Plenteous will be the reward; maybe even a luxurious fur.

Don't hide the ignorance you bear.

Embrace it, then focus, maybe glare.

The heart of the matter will gain worth, enlarge, and become fatter.

It will fuel so many;

Not simply with money, a dollar, or penny;

Enrichment is experienced each moment.

Springs of sparkling wine bubble then ferment;

Tantalizing the senses, and eventually making a heightened awareness.

Jealousy will stab you like daggers.

Others will view you with disdain.

Don't argue, create refutes or disputes.

Be careful to avoid and refrain.

THE LOW AND THE MIGHTY

Some are quiet, others sad;

Many in disparaging tones are called mad.

Within society varieties lie,

Containing cultures, traditions, beliefs;

But where has gone the value of a single leaf?

Where people do as they please,

Invariable,

A portion will fall to their knees.

Question why it is a person will see and acknowledge wrong,

But still decides to be among the throng.

Do pangs of guilt inflict his conscience?

Or has this become a matter thought as nonsense?

Without a doubt, man collectively has changed;

By not paying a price, bearing a consequence,

He loses touch with himself and others,

And avoids being pained.

Insensitivity, sluggishness of thought, out of tune;

However,

Consumed with things that can be bought.

On and on, issues, conflicts, arise,

Each one thought deserving of being fought.

A loss of tranquility is the price.

Clear mindedness, level headedness,

Become as calculated as the throw dice.

Giants were once seen in utter awe;

Sources of inspiration, and so much more.

Now their words are neglected,

Maybe this is due to individuals specializing,

Focusing solely on one thing.

We lose awareness of a developing, heightening, frailty.

We cease to care for others,

Thus, obviously, for oneself.

Books of wisdom and grand knowledge

Hidden within tales

Are left gathering dusty on a grimy shelf.

STANDING THE TEST OF TIME

So many of my dreams have vanished in the wind,

Never again likely to be seen.

Hope springeth from the child's chuckles of delight.

Simple things, like the love of flight,

A kite swaying to and fro, among clouds,

And being given feathered wings,

Seeking to reach a soaring new height;

Held by the tender hand,

Becoming increasingly acquainted with the worth of

The power to know.

A job of significance and worth;

How splendid to make a contribution,

In an evolving world full of hideous things,

Not fit for a child's senses to behold.

Where is the chance to take a firm stance?

When around is the love and adoration of gold.

Environment creates, manifests, alack of thought.

Never mind, all will be well, people say,

As long as a gadget is available, and can be bought.

The irony of it all

Is that so many sense an economic crisis

When nothing could be further from the truth;

Enrichment of the self is actually our call.

To stand against all this

In a darkened cesspool, some call a mist,

Is ineptitude, waste, incompetence, ignorance,

An overall lack of morality,

Which excuses the present

Dire and desperate states of inequality.

The message to take note,

Is that man be design or hideous fate

Is responsible for this abominable folly!

GIGANTIC FOOTPRINTS
ARE WITHIN
YOU

Watch the leaf as it falls to the ground,

It reveals its front and back;

If tasty, it may be eaten by a hound.

Our destiny is obscure.

We have no way of knowing where and when the end will be;

A road can be rocky, yet offer us all we need to see.

Though weary, pastures offer pleasure,

Horses grace, cows rest in states of tranquility.

Heightened perception reveals within each an immeasurable

Sublime treasure.

Life is often not as we would like it to be,

Troubles, problems, spread, disperse, collect, dissipate,

Vanish then reappear;

All the while believe in yourself:

Contained within is all that is required to be complete.

White and black, various shades lie between.

We are not even, but gradients of impurity.

Accept this with pronounced sincerity.

With time, effort, endurance, perseverance,

Revealed will be compassion enormous, herculean.

David, is how you appear in the present,

A Goliath of virtue and growth will be the end result.

NON THE WISER

I am lonely

Because I sense no one is there.

Hidden faces; the truth lies buried within.

Where is the

Sincerity, truthfulness, and honesty, I so desperately need?

These are the things on which I most eagerly feed.

Our true nature is contained within all these qualities.

Observe each; present are our frailties.

They are not weaknesses,

But elements that surely and must grow.

A field manufactures wealth,

By using a heavy, durable, penetrating, hoe.

Doing what is right, often causes pain.

Each one, regardless, must endure this passage;

In the end it will be realized,

We are all one and the same.

TRUE WORK
LEADS TO TRUTH

I've grown weary, tiredness fills my soul;

Where is it I now should go?

Around the corner lies a hopeful dream.

Disappointments, frustration galore,

Have made me question what it is all for.

To make better oneself, is the key goal,

Yet today so many choose to live on the dole.

They care not about others, obviously, also not themselves;

Ignorance is their bliss;

Because they are afraid of what they might see.

I haven't a clue why a man would choose such a disorder.

Gold, jewelry, fur, luxuries, are their fodder.

Content they will never be with what they've got;

Measuring, equaling, becoming better than others,

Makes them excited,

Dare I say, hot!

Where is the quality of man among these desires?

An ego is what hampers a man from becoming his best.

Invite others into your home;

Let them feel the pleasure of being your guest.

Before you know it,

You'll arrive upon the true path that is your honest quest.

Judging others will eventually bring us all down.

Listen instead to your heart,

Then sooner than you can possible imagine,

It will be harmonious, sound.

Creatures fueled by passions, lust, and anger,

Search above,

Reach toward the one deserving the title as our true founder.

A HAND FULL OF
DESIRE

I lie beside lions as they sleep and graze;

Meanwhile, I can hardly manage to withstand this phase.

I see shadows in their eyes;

No matter what I perceive, each one of them lies.

Tough, fair, handsome, robust, apparently sane,

Inside lies their hellish destiny,

None of us would like to face,

Or deal with the fact that toward that they proceed;

Down they go, no longer having a hope they can grow;

To the beginning instead, the time life was a seed.

A smile, a handshake, a caring gesture,

Is all they can manage to provide.

Limited in their capacity to cope;

They lean increasingly toward that, which is dark,

And the others that reside on their side.

Look closely, before you is the blatant truth;

A fur, a Mercedes, a home large enough to fill an ark:

Represented is a small, minuscule, mind.

Undeveloped is their heart, mind, and soul.

All that is good and our best;

Which is something that should never be settled for anything less.

Right, proper, equal strong,

Means nothing to the members of this throng.

They devote time, effort, gold, all to themselves,

While the victims of theft, harm, and shame,

Are kept in cages, jails, commonly called cells.

The meek and the humble shall inherit the earth.

They care not about being first;

For knowledge alone they thirst.

The road is long, rocky, hazardous and dangerous.

Forget all this, focus instead on the self.

Complete, whole, unique, quaint, fine, thoughtful;

What is it still you are worried about?

DAYS OF FUTURE PAST

I can see further, past the horizon,

To a land where there is still bison.

Land locked times of hope,

Where creatures were lassoed with rope.

Further back in time,

Man was more in touch with nature.

He lived to know about rocks,

Comforts were often scarce,

Entertainment wasn't derived from fairs.

Self created was his world.

Pleasure, satisfaction, nutrients of life and company,

The multitude varieties to capture his mind,

Were fashioned by his endearing hand.

Coarse, rude, some what muddled and confused.

He would be seen by the eyes of many today;

But think how little is required to make the masses shrink and fray.

Our world today contains hosts deemed strong,

But lands, regions, and populations, have fragmented.

Decay, despair, inflict the poverty stricken,

Tied they are to the mighty who hold riches.

Greetings are given by their beautiful bitches.

Lies fuel fantasies that deny others hope.

One day around their accusers, persecutors, will be a rope.

Only a minute portion of the human race will live to truly survive.

Possessions will entail all that they desire to thrive.

A SPECIAL SURVIVAL

How does survival happen?

By chance, or because we take a stance?

Why drive to continue,

When so much around we see as pitiful and sinful?

Ignorance abounds

And hounds those who hold knowledge dear.

Their peace is provided by an allowance,

Granted by the one that offers splendors;

A blissful state often seen in dance.

Guardians of life, they appear to some,

Others, a dreadful, despicable, bum;

Regardless, they are the ones that hum,

Having seen and understood things referred to as

The One.

They are the ones truly free.

Behold, there are rings of fire encircling their dear, special, frames.

Life, they have realized,

Is a series of splendid, ingenious, games. |

They have wings not seen,

Carried on pillows that resides inside their mind;

Severed they are to the ties that blind.

Gentle is their healing touch; often they say to others, hush.

Worry not, all will be well; after all, there is no such thing as hell.

The answers you require were not spoken by a burning bush.

Fear not.

Everything you need, you've got.

Hold tight,

The ride we are on is long, treacherous in so many ways.

Is this our dispensation for all our days?

Am I the prey?

Where are the precious things that jeopardize things?

Forever, I seem to be running from some hunter.

Of course, this could not be the case.

We have been given truths;

Used carefully, they will serve as our base.

We have also been allotted the darkness that falls at night.

Rest well during this time,

Then one day you will reach your true height.

At night, those who are wise, pray.

Life, we know, often appears grey.

Black and white,

Have many shades that lie between.

These are often the hideous sights, and nightmarish visions, seen.

The absolutes are veiled.

There remain too many whose quest is merely to get laid.

Murky are so many things, because people are simple,

The greatest care is the sight of a pimple.

At the end of each day,

Ask that dangers be kept at bay,

Because we wish to stay

On this green earth we have here with us today.

CREATIVITY

Creativity, the art of building.

From things small,

We collect till there stands something tall.

Structures of thought, ideas, deeds, pent up feelings;

It is with these things we are dealing.

To be an originator,

You must first be prepared to stand outside the box.

Society restrains us from telling tales

That happiness can be derived from eating meals

Consisting of cream cheese and lox.

Artists require much more than the basic necessities of life.

Everyday a worlds they create.

So often they are alone;

They cannot find a suitable mate.

Difference, is the key that opens the door to sights previously unseen,

And where we'd like to

Find ourselves to be.

So often these Creators aren't appreciated,

Instead of being seen as Giants,

They are small, as inconsequential as a flee.

How unfair it is that greatness often isn't recognize for its worth.

We've reached a day when ignorance of this kind

We should dare not stand;

Thus, we require a new start, a rebirth.

If only the kind and mighty of mind are left behind,

Maybe we will have a chance.

Once again the earth we will appreciate.

Flowers will offer pleasure,

And songs will be sung.

Along with each we will gladly skip and dance.

Times have changed.

Many things have evolved

To the point where so few wish to get involved.

On crowded streets people walk and talk,

Yet still a distance is apparent.

People have to have a caring parent,

Or a meaning and substance is not there.

Thus, despite what is seen,

These people think little of those around.

The truth is they really don't care!

MAN

What makes a man?

Many things of a kind we do not understand.

Days pass, we are awake,

And when we close our eyes, we sleep.

Through it all we have a journey we must follow.

The answer to it all, we always hope, will happen in the morrow.

We search, we follow,

Until we find a shore on solid ground.

We can call our own.

Some call this our final resting place,

Others, home.

No matter how far, or how fast, we travel,

No time is passed.

We will always end up where we arte now.

The beginning is the end.

The question many ask is;

How?

Trying to understand all this

Can sometimes send us around the bend;

But still we pursue the quest to understand.

Though, from one day to the next,

We appear no closer to the truth,

To the Promised Land;

Is this in fact true?

Or do we need a helping hand?

The search without will make us weary;

Finally, we decide to search within.

We develop a new theory;

The ultimate is obtainable:

That is, true knowledge of everything that has been,

And everything we could p[possibly see.

The complexity of us, is daunting to comprehend.

Layers, stages, rungs, stepping stones;

The essence of our being goes deep into our very bones.

Then, we say,

Everything outside, is a reflection of what we have within.

What is it we avoid most? Is it fear?

It inhabits unknown territories, guarded by a shadowy host.

We fear ourselves;

Can we stand the truth?

Or will we fall,

And maybe, possibly, never rise again?

Oh, the pain.

Existence is merely a game.

Let's end it!

Free ourselves from all suffering,

For all time!

The Buddha said it all.

Many Giants there have been,

But he stands as the tallest!

The others, in comparison, are quite small.

TRUTH

The truth stands before you like a shadow in the night.

It isn't hidden, it is light.

Hollering out your name,

Asking for attention,

But we choose not to heed this call for salvation;

Instead we devote ourselves to a nation.

It is merely a land, where so many souls lie damned.

Today we stand as fools,

Believing happiness can be derived from having pools.

Dive into the ocean of life,

That is what we should do;

Only then will we discover who named who.

It wasn't you!

Then who was it?

Many times the answer's been given throughout time.

Think hard, believe in yourself.

The answer will rise like the dawn.

It leaps forth like a loving, graceful, fawn.

God!

That is who!

Now you've got something to hold.

Doesn't it make you bold?!

It's the truth.

Now you have all you need

To know those things that have been said

Are true!

AFTERWORD

The Greatest Lie Satan Ever Told Was Making Man Believe He Doesn't Exist

Jesus Christ, The Preserver part of the Holy Trinity, is reported to have said; "Suffer the little children who come unto me." One can then conclude that somehow within the scheme of things suffering is good, and keeping yourself much like a little child is something of benefit to you as well; which most, I'm quite sure, consider contrary to their reasoning in regard to how they may obtain The Lord's approval, acceptance - some call this, grace.

The meaning of the above expression used by Jesus can be understood more fully by examining the encounter that occurs between the Serpent, representative of Satan, temptation, the destructive force – otherwise called the Holy Ghost. We are actually being informed during this event how the conclusion of this Age will come about.

Within the Garden of Eden Adam and Eve are given just one instruction, some might call it, the solitary Commandment; do not partake of the fruit stemming from The Tree of Knowledge of Good and Evil.

The Serpent is aware that Eve is drawn by touch and sight to the apple, and tells Eve that The Lord is withholding information from Eve that He doesn't want her to know about; namely, if she does eat the apple she will be equal to The Lord in every regard. The temptation must be enormous at this juncture, because if such a thing is in fact true, Eve will not pay a price, bear any consequence, due to the act, or anything thereafter.

There is a fault in her reasoning that is easy to distinguish; The Lord created her, she is The Lord's creation, therefore, she is from The Lord Himself. If she were to become equal in stature, there would be two Lords, but she is a part of the one that created her, not separate; when Adam doesn't question where the apple he is presented came from, and then decides to eat it, both Adam and Eve are evicted from the Garden Of Eden.

This is because they have gone against their nature, how the universe operates, the laws it obeys, which is what The Lord happens to be. Jesus is reported to have said that a person cannot have two masters, and if such is the case, the person will inevitably fall.

Outside the Garden of Eden there are trials, troubles, and tribulations, some might simply refer to it all as, suffering. This is not a penalty that has befallen them, but something Adam and Eve imposed on themselves.

Surveying the earth as it exists today, it doesn't take much to recognize that a massive portion of the world's population deal with enormous hardship practically every day of their lives; obtaining the mere basic necessities of life in their primary occupation: considering there is far from a shortage of such materials in the world, the suffering man has placed on his fellow man is not necessary, but an illustration of the extent to which he is operating against his nature.

Eve's frailty was her inability to adequately question the validity of what the Serpent was saying; her cognitive skills were weak, which can also be described as having a limited power of will and a lack of awareness, which resulted in her not seeing through the lies she was being told.

Suffering is something none of us, I'm sure, wish for ourselves, and we do our utmost to avoid at all times; therefore, if Eve is able to accurately decipher the weakness within herself, and take the necessary measures to rectify it, the inevitable result will be a return to her prior state which existed in the Garden of Eden. Had she been able to correctly question the tale told to her by the Serpent, she would have realized the claim being made was impossible.

Little children, who are brought up well, know right from wrong, and are not inclined to question the obvious, thereby wasting precious time which could otherwise be used to satisfy their eager curious minds. As they explore, mistakes are made, for which they suffer, but then learn as a consequence to avoid repeating in the future.

When they encounter others not inclined to behave in the same manner, predictably a conflict of interest will arise. The child is faced with something that can harm him, which he then takes the required measures to avoid in the future.

We live in a world where both men and women do not sufficiently question what they are being told. Tragically, many suffer as a consequence, but no penalty, price, punishment, is attached to decisions that result in the loss of human life, which, thus, deprives the perpetrator of the chance to suffer and learn from their wrong acts; such a pattern weakens their faculties of free choice and free will, which could be called, their soul.

The only end result that is possible is that eventually people will become completely clueless, in a manner of speaking; they will have no suspicion of the damage, destruction, despair, and misery, they cause - not just others, but themselves.

When the point in time arrives that their soul has gone, they will not have an awareness of this, primarily due to depriving themselves of the faculty that enables such a thing to be possible.

The greatest lie Satan ever told was making man believe he doesn't exist. Satan is any person masquerading as a human that does not have a soul. They become this way due to incorporating within themselves substances that are foreign to their nature.

A common expression is, "you are the company you keep." A person becomes much like the environment he chooses to inhabit and engage with; being around liars makes a person prone to become a liar; till, eventually, the larger the lie, the easier it is for the person to believe.

The monsters humanity fears, are the same as what they are. The United Nations has declared that we all have Human Rights, when any Right is violated by any means, to any degree, you are being attacked, and your soul is placed in jeopardy.

The Lord, obviously, cannot change who and what He is, but by viewing the actions undertaken by Adam and Eve, and the population of the earth that exists today, He can see the scope of what He is, and at the same time be less alone, because the suffering that results from a person separating himself from his Godlike nature, is a reminder to Himself of what He is.

After all is said and done, and no part of humanity is able to return to the Garden of Eden, The Lord then has the opportunity to be alone, and reflect on Himself as it has been revealed within the book of life, which he created Himself at the beginning of time.

Man Be Kind.
The Lord is King

Heaven will be on earth

when man realizes he is the same as God.

"Know that you are all gods, and the
kingdom of God lies within you."

The Father, the Son, the Holy Spirit